Schpleee Technologies, Inc.
System for Expedited Vaccine Inoculation

OR

The fastest way to Herd Immunity, And How to Rid the World of COVID-19

By Curtis Raymond Crim

Dedication

This book is dedicated to Vice
President Kamala Harris
For inspiring generations of young
women –
The future of the Human race.

TABLE OF CONTENTS

Introduction

Human life. That is the bottom line. There are many considerations in a massive effort to inoculate our entire population using modern technology in the current state of the economy. Efficient use of the vaccine available, environmental impact, and cost considerations are among them. I address these issues, but I submit to you that there is *really* only one important issue: human lives. Each life is special, loved, and irreplaceable. The faster this (and any other) viral pandemic is cured; the more human lives can be saved. This document is written from the point of view that *human life and the saving thereof is the only important factor.*

To the person who sincerely wants to cure our country and the world of the COVID-19 virus (or any other viral pandemic) a reasonable approach would be to use the current infrastructure and system immediately available. This approach, although old-fashioned and obvious, will be moderately effective, but it is simply not enough. Due to the exponential nature of the spreading of a virus, the more quickly it is eradicated from the world, the more lives will be saved.

Speed of distribution, therefore, becomes the key to getting the pandemic under control and saving as many people as possible in the shortest period of time possible. To accomplish this, I am separating my system into four broad categories which are (abbreviated): Education, Distribution, Database/Web Interface, and Globalization.

Note that I am not addressing manufacturing at all. This is for several reasons. One is that I do not work for a vaccine manufacturer, and have no access to their systems. Further, I submit that eventually, the manufacture of adequate amounts of the vaccine will inevitably become a self-resolving issue. We are now facing at least three major "bottlenecks" in terms of inoculating the entire country (and the world for that matter). Manufacturing vaccine is one of them (obviously), but in time there will be plenty of vaccines being manufactured as the demand for the vaccine will continue to diminish. The outcome will be that the virus is gone or mostly gone, and the ability to produce massive quantities of the vaccine will have become trivial.

This document focuses on the process starting from a manufactured vaccine and ending with the achievement of herd immunity, and hopefully as close to 100% of the population inoculated as possible.

A big part of the problem with addressing a pandemic of this magnitude is that we have a weak cultural context for dealing with the issue. The world has not experienced a pandemic this extreme and widespread since 1918. The computer was not even invented at that time, and the culture that exists in the world today is very unlike the way it was the last time this kind of disaster took place.

I suggest that for a society to survive, it must evolve and adapt to environmental conditions (including the presence of a pandemic virus). In other words, to survive this situation while minimizing the loss of human life, we must as a culture evolve and change and adapt in order to survive in this new version of the world in which we live.

I believe that the key to changing our culture itself to meet and defeat this pandemic will (and should) start with education and an adjustment of our culture's attitude.

The Modularization of the Schpleee Technologies, Inc. System For Expedited Inoculation

One major advantage of my system is that it is easily divided into parts, and that any one part can be implemented without the others. If a country or municipality desires to implement "Phase One" and none of the others, it will still contribute in a major way to defeating pandemics in general.

Further, my system is totally copacetic and compatible with all other systems for inoculating a population currently being used, either in series or in parallel. This is not an "either/or" situation. This system is not mutually exclusive with other systems or endeavors. It works perfectly in concert with all efforts that are currently being implemented.

Other advantages are that implementing my system is potentially fast, it functions amazingly quickly, it is low-cost, **and** it is environmentally friendly (details to follow in this document).

Phase One: Education

Part One: Attitude

For those who are now wondering how attitude can help defeat a pandemic, please allow me to explain. I sincerely believe that to survive this pandemic disaster and to be ready to eradicate similar disasters in the future, we must adapt and change our culture. This is not 100% essential, but doing so will increase the speed and efficiency in defeating this and future viruses, and could potentially save the lives of millions of people over time.

Every human life is precious. Even the loss of one single life is terrible, as something unique and special is erased from the Earth forever. My motto is: "Not one more life lost". Obviously, that is not reasonably possible, but I want to see people of all nations come together to minimize the loss of human life. Most people, I think, will agree that this is desirable.

This world as it stands is different from any we have seen and existed in previously. To survive, we must adapt. To adapt, we have to change our culture itself to become a metaphorical living entity that is focused on protecting itself from the COVID-19 virus and any other viruses that might come into the world in the future.

My aim in this chapter is to provide suggestions as to how to integrate a positive attitude, in terms of defeating a viral pandemic, into our culture itself. The way out culture appears, the way it functions, the experience of it, and the way we implement it.

How we *see* our culture, is germane to this endeavor.

The implementation of my system for defeating the current and all upcoming viruses starts with how we as a culture see ourselves. It is not simply inoculating individuals to get the current situation under control. It is about defeating the next pandemic *NOW* before it even gets started. Our survival in my opinion depends on changing the way we see our culture and ourselves, and also our family units.

To change the culture, *attitude is everything*.

I am saying that we need a new vision of whom and what we are as a people. I am suggesting that our cultural self-image must change first. I am a scientist and an engineer, so this document will focus heavily on technical issues, but to start with we must first focus on *cultural* issues.

Let's start by discussing historical American self-image. Some of the concepts in cultural identity in the USA include hot dogs, baseball, loving your mother and country, and apple pie. We sometimes refer to ourselves as "America, the land of the free and the brave".

This is all great, but I would like to expand and embellish this cultural self-image. Here are some of my suggestions for new mottos that incorporate the adaptation to the presence of pandemic viruses and the defeating thereof: "America the strong", "America the Healthy", "America: Virus-Free", and "America the Pure". Permutations and variations of these mottos are perfectly acceptable. I am suggesting that we modify the culture and perception of the culture of the USA. Our cultural identity itself must change.

This chapter is actually focusing on education, which has not yet been discussed. Moving toward the discussion of education, we should look at how we are to present the issue to students of all ages, especially the young. I envision the future of the youth in America learning of their country as "America, the brave, the healthy, the pure, and the virus-free."

This might seem to be taking things too far, but I submit that it is not. I think that the situation is far direr than most people want to think or feel or admit. We must change and adapt for the survival of the human race.

Changing our collective cultural self-image is a small price to pay to accomplish the goal of saving as many lives as possible and possibly even saving the existence of the human species.

Incorporation of an Additional Discipline

Into Our Culture

There are many options for a community and/or culture from which to choose to teach people to give injections to each other. In this section, I am going to suggest a few, but it is also upon each community to choose how to accomplish this goal.

In-person training: I believe that communities can set up local venues to offer courses in the training that I suggest strongly that all individuals in a culture receive. These local venues, among others, might include community centers, senior centers, churches, law enforcement centers, fire-fighting facilities, and of course medical centers. Use your imagination. There are a virtually infinite number of possibilities.

Training Incorporated into Education Systems: Cultural leaders and government law-makers will have to make a call as to who qualifies for certification for administering intramuscular injections. I submit that individuals who are very young, new-born babies, for example or those who are very old (to the point where mental senility prevents them from functioning) will not be capable of administering such injections.

Excluding those who are not physically or mentally capable of giving another individual an intramuscular injection, I believe that at least 80% of any given population will still be able to provide this service.

In other words, MOST individuals (perhaps over 90%) in any population should be able to give someone else a "shot" effectively. So, for argument, I am going to assume that humans as young as perhaps 10 to 13 years of age will have the physical and mental capacity to understand what they are doing and be able to administer an inoculation.

Therefore, perhaps as early as in the 5th grade, in the USA, we can start training and certifying potential Vaccine Inoculation Nurses. To make this practical financially and expediently, the already-paid school nurse could hold a one-hour assembly once a year to teach or reinforce the method for giving injections. This will hold the cost and time involved to a minimum.

The goal of this phase of the system is to address one of the major bottlenecks in getting our population to Herd Immunity.

The fact is that during this pandemic, most or all of the doctors and nurses who have somehow managed to survive so far are desperately overloaded in terms of the work that is required of them. They have plenty to do and don't need the additional task of performing inoculations adding to their terrible burdens of caring for those who are already sick.

What we need is a whole new work-force available to perform this service without burdening the already-strained medical community.

Therefore, we need to train a whole new generation of nurses, wherein almost every individual can administer an inoculation injection. This might appear trivial right now, but it won't in the future.

Using this system will remove this potential bottle-neck not just for now, but for all times. For any and every pandemic that comes along, we will be ready. America will become an army of vaccination nurses, ready to move in an instant whenever the need calls.

I believe that this can be incorporated into our education systems at a minimum of cost. This skill is very simple. It is far easier than driving a car, and I think that anyone who can get a driver's license can also be certified to administer an intramuscular injection. So, at about the age of 15 (if not before), people can become certified to give injections in most states.

Web-Based Training: The most obvious use of modern technology and perhaps the most powerful tool in this arsenal will be to allow on-line training of Vaccination Injection Nurses (VIN's).

One suggestion for a URL is: "covid19US.gov". The web site will include a video-based training course to be followed by a final exam, the passing of which will allow the newly certified vaccination nurse to legally administer antivirus vaccine, print out a nice frameable certificate that they have the option of mounting and displaying on a wall, and most importantly, order vaccine as needed.

This web site will allow millions of people to become certified VIN's quickly and supplement, in a substantial way, the population of available individuals capable of administering the life-saving vaccine that will eventually defeat the pandemic.

This section also deserves some kind of a motto. Here is one possibility: "Give a vaccine, save a life. Teach someone to give vaccines, and save many."

Goals

I am not comfortable telling others what to do. It goes against my morals. Even if not stated outright, take every command I appear to give as a suggestion or a recommendation.

The first (suggested) goal of this system for ridding the world of the current, as well as all future viruses is that every household in America should have at least one resident who is a certified VIN. Eventually, the long-term goal will be to have EVERY member of EVERY household in the USA a certified VIN, as much as is reasonably possible, in terms of who is in an acceptable age range, and taking into account physical or mental infirmity that would prevent them from being capable.

This idea is crucial because it addresses another bottle-neck that is currently a concern and problem: There are not enough inoculation/vaccination sites available in many areas.

Making not just a hundred thousand new certified administers of inoculations, but potentially hundreds of millions of individuals in our culture both trained and certified in giving intramuscular vaccination inoculations removes one bottle-neck for all time. Making (and this is a KEY point!) *EVERY HOME IN AMERICA AN INOCULATION SITE w*ill remove one more bottle-neck for all time.

I hope that all who are reading this text will realize the various advantages of this part of the system. In an emergency, a person could potentially walk up to almost any home in America and be able to receive a vaccine inoculation protecting them from whatever pandemics are prevalent in the world at the time.

It will be an enormous advantage for one single VIN to inoculate and protect the members of their immediate household, but it is no stretch to understand that the person dispensing vaccinations can also inoculate the home next door. This is a great advantage, because it propagates multiplicatively.

A VIN will also have the ability to *teach* vaccination technique to any person whom they inoculate. The ability to administer a vaccine needs to spread exponentially, similarly to the way the vaccine spreads.

If one superimposes a graph of the ability to distribute and vaccinate on top of a graph of the exponential spread of the virus, one can see several possible scenarios. The line representing COVID-19 number of cases is demonstrating an exponential curve up. The second line represents our ability to distribute and apply vaccines. Its behavior could follow these possibilities:

The flat line: Actually, this line is going down when nurses and doctors die from being infected by the virus they are fighting. This model assumes that we maintain a steady number of (overworked) technicians/nurses and doctors available. It takes years for a nurse to graduate and become certified. It takes years more (up to 10 or 12) for a doctor to do the same. If these are the only people who can administer the vaccination, it will result in the longest period of time to achieve herd immunity and also result in the greatest cost in human life.

The straight inclining diagonal: If we can certify and create new VIN's at a steady pace, this line will overtake the exponential spread of the disease more quickly but the cost in human lives can be reduced further.

To get the advantage against an "enemy", the COVID-19 virus, which can spread exponentially, we also need to get an exponentially expanding distribution system to give us the best chance to defeat the virus more quickly and thereby save the maximum number of human lives.

Propagating One Simple Skill

Some might object to the idea of training people in general in a skill that is usually associated with the skill-set of an individual in the medical industry. It is time to let go of this old concept. Giving an intramuscular injection is amazingly easy, and most human beings can perform this task.

To reply to the aforementioned objection, I submit that we already have a cultural precedent as to why this is a reasonable course of action.

One example is the EpiPen. This device is an accepted part of our culture, and is often carried by people with severe allergies.

I personally have had allergies and asthma my entire life. When I was a teenager, my doctor, Conlon, allowed me to take my injections home and apply them to myself rather than having to go into the doctor's office.

Many diabetics have to learn to give themselves injections. I believe that this is actually a different kind of injection, but it is not a far stretch to go from this skill to be able to perform an intramuscular injection.

My vet has allowed me to bring my pets' shots home and apply them myself. This is also a different kind of inoculation, but again, it is not hard to give various kinds of shots once one becomes accustomed to doing so.

The point here is that our culture already has several contexts in which it is considered normal for a person, who is not medical personnel, to administer inoculations. It is not hard to do. Given that we are currently in an emergency, I feel that it is acceptable to disregard the objections to not using actual accredited nurses and doctors to perform this service.

I have noticed that President Biden has also suggested the use of "non-medical professionals" to give vaccinations. I think that this is a wonderful idea. My system takes this idea to the next level.

Ecologically Friendly

Another advantage of implementing my system is that it is amazingly ecologically friendly, and will reduce the carbon footprint of inoculating a population.

Consider this scenario: A family is staying inside and observing proper quarantine protocols, regardless of comfort. Are the individuals in this household expected to drive to a vaccination site and wait in line to get their shots?

I certainly hope not! Just imagine the gas being burned. This approach makes sense in terms of the old way of thinking, but my system offers a superior approach.

Why ask a family to break quarantine when it is not necessary? Any capable adult member of the family can log onto the governmental VIN certification web site and become the person who will take care of the entire family. The web site should also provide the option of ordering vaccinations to be sent directly to the household from the manufacturer.

After the no-contact delivery of the vaccine, the household's VIN will administer the vaccination to the entire family, *without risking lives, and without burning any gas.*

Further, mass-vaccination sites could potentially become super-spreader venues. Having people car-pool to reduce carbon emissions only increases the chances of transmission of the disease.

Also, consider that the Schpleee Technologies, Inc. solution is low cost.

Medical personnel will not have to be paid to perform this service.

The Schpleee Technologies, Inc. system will be safer than most currently used systems. It is not only safer but also low-cost and environmentally friendly, reducing the carbon footprint involved in inoculating a nation's population.

This chapter has focused on the education and certification of a new generation and army of Vaccination Inoculation Nurses. The next chapter will address getting the vaccine from the manufacturer to the individuals who will administer the injections.

My Vision for Educating the
Entire World
To Give Intramuscular Vaccinations
In About One Day

President Biden announces a State of the Union Address (of sorts) in the AM of a chilly day in February.

Later that day, the evening is cool; the crowd silent with the electricity and excitement and anticipation...

President Biden walks out and the crowd roars! After a great while, he can get control of the crowd, "Thank you for coming to tonight's Presidential Address." Another roar is heard from the crowd.

"Tonight", he goes on, "I am asking every American citizen and those all over the world to learn how to administer an intramuscular injection. I am asking each one of you for just 10 minutes of your time to do your part to help defeat the COVID-19 pandemic. Stay tuned after this educational video for a continuation of this live broadcast."

A government-approved instructional video is displayed over

live audio and video feed that teaches step by step how to prepare a syringe for the inoculation process, how to apply antiseptic to the inoculation site on the patient, remove the syringe and apply a bandage to the site of the inoculation.

When the educational video ends, live feed pans to a nearby stage with President Biden and Dr. Fauci, with several other important people standing in witness, including Vice President Harris. Dr. Fauci then repeats the process just demonstrated on the video live and teaches President Biden to administer the vaccine to a patient, which the President does live and on-air streaming on the internet and News networks world-wide.

As the President removes the needle and applies the bandage, the crowd goes crazy. President Biden turns to the crowd and cameras and says, "If I can learn to do it, you can learn to do it too. If Dr. Fauci can teach me, you can also learn to teach others."

President Biden returns to the podium as the crowd cheers and continues to address the crowd. "Tonight, I am setting for America the goal of having at least one member of every household with the ability to administer an intramuscular injection. We are launching a government web site that will provide step by step instructions as to how to administer the COVID vaccine so that we will be prepared for the next step in defeating this virus." He goes on, "Every household in the USA should be prepared to receive enough vaccines to inoculate every member of your household based on the 2020 National Census.

The President concludes with appropriate inspiring remarks.

The live video ends.

This event will not only demonstrate how to administer a vaccination shot but also how to teach others the technique; this live event will be witnessed by at least hundreds of millions of people. I would give it a day to propagate and go viral, and another 24 hours to propagate worldwide. Most of the modernly industrialized world will be able to perform the needed injections within a couple of days. Perhaps over fifty percent of the world's population will be adequately educated in giving these life-giving shots in about a day, using modern available technology.

People tend to take their leader's behavior as an example as to how to behave. If a president is intelligent, mature, morally upright, decent, and considerate, then it is likely that his followers will behave with conscience, respect, and wisdom. The people will want to imitate the President's behavior, which demonstrates the appropriate, responsible, and moral behavior desired from every citizen.

If on the other hand, a President were to behave in an illegal and immoral manner, it is likely that his followers would similarly demonstrate like behavior.

One advantage that this idea has is that the President can do it immediately with almost no cost and no approval from the legislature. He can address the union and say whatever he wishes to say, and needs no one's approval or permission.

The best part is that a big part of this system, educating everyone in how to give an intramuscular vaccination inoculation will be done virtually overnight.

The people of the world will be prepared for Phase Two.

Phase Two: Distribution and New Inoculation Technologies

In an emergency, always develop options. Speed is of the essence. Every second you save can equate to the saving of many additional lives. This is especially true when hundreds of millions of individuals are involved. *Every second* counts*!*

By streamlining the process required to administer an injection, we can save seconds or even minutes per individual, and thereby multiply the number of people we can get vaccinated over a specific period of time.

I applaud the systems currently in use that are now vaccinating millions of people. I suggest there are obvious ways in which these massive inoculation efforts can be decreased in terms of the amount of time required to vaccinate one individual.

For example, why should it require the time of a trained nurse to apply the bandage post-inoculation? Most humans can master this task at the age of approximately five years of age. For those too young to master a bandage, their parents can apply it for them. Either way, this task can be passed along to those who need no related professional skill set and not being paid for their time and effort. I suggest that a dispenser or even a simple bowl or box be provided at the venue exit, with instructions to take a bandage if needed. The public can also be advised to voluntarily bring a bandage. As I will continue to repeat, always maximize the number of options provided.

To further streamline a mass vaccination venue, a dispenser with antiseptic wipes could be provided such that the person being inoculated would be instructed to bare their arm and then wipe their desired inoculation area, after which the vaccination nurse would apply the vaccine, and they could then acquire a bandage as they exit (and then optionally apply it personally).

My last suggestion is to have a 2-person nurse crew. One person will just prepare syringes and hand them to the other nurse, whose sole job would ideally be to just receive a syringe, apply the vaccine, and throw the syringe in the trash (and repeat). By isolating sub-tasks, the overall speed of the operation could be multiplied.

If even one minute could be shaved off of the average time involved in inoculating one individual, then 1,666,667 man-hours could potentially be saved, every one of which could be used to inoculate more individuals, rather than throwing it away having trained nurses and doctors doing non-skill requiring tasks like wiping an inoculation site or applying a bandage, both tasks that almost *anyone* can perform! If an average of TWO minutes could be shaved off of the time required to inoculate a single individual, then over 3 MILLION man-hours could be freed up to inoculate additional individuals in a population. That means that by speeding up mass inoculation sites with these suggestions, millions of additional lives can be saved.

Note: Every one of my systems depends upon sending the vaccine directly to the people who will be receiving them, and/or the nurses who will be administering them. Frequently, these will turn out to be the same people. Also, a key element of both systems is spreading the ability to give vaccinations and teach others to do the same at least as quickly as individuals are being vaccinated.

Also please note: None of these systems are reasonably compatible with the Pfizer product, or any that require extreme storage or transportation environments. However, the Johnson & Johnson vaccine would be *perfect* in terms of compatibility with my systems. AstraZeneca would also be a great candidate. The single-shot scenario is far superior if you have individuals in the population giving the vaccine to their own persons; the less stringent requirements for storage environment temperature will be more easily accommodated by anyone with a working refrigerator, which will include almost every household in America.

I will be discussing two different systems for distributing the COVID-19 vaccine at a highly accelerated rate.

Distribution System One:

The Fastest Way Possible

The basic idea with this system is to just ship COVID-19 vaccine directly to every household in America based on the 2020 census. The census guarantees that the government of the USA has a database that contains a majority of resident addresses in the country, as well as how many people reside at each address.

My suggestion is that a small kit be sent to each address with not only enough vaccine to inoculate each resident of a particular household, but one that also contains a sealed alcohol swab or tissue packet and a bandage for each resident. The vaccination kit will also include an ice pack to keep the vaccine fresh.

This kit should include a full-color pamphlet with easy step-by-step fully illustrated color instructions as to how to perform the inoculation (for those with no Wi-Fi internet access), and the URL of the government web site (suggested: http://www.covid19US.gov/cv19vaccine_instructions/) which they can access using any cell phone, tablet, or desktop computer if they do have access to the internet. This web site will offer a free tutorial as to how to apply an intramuscular vaccine injection, among other useful resources related to the COVID-19 pandemic and achieving herd immunity for our population, to help to rid the USA of this terrible virus completely and for all time.

The American Home Antivirus Vaccine Kit can also be enhanced with the development of one more great option for individuals in our population that would provide greater ease in the vaccination process itself. It would require a small amount of development effort, but it would be well worth it.

The new item to include in the kit would be an EpiPen Auto-Injector style of a syringe, one for each household member, which we will now christen the "COVID-19 Pen" or "CoviPen". This kind of intra-muscular syringe was designed to make it easy for almost anyone to administer an emergency injection to a patient, including self-injection. By sending pre-prepared easy-injection style syringes, it will far simplify the process of applying the inoculation by the average citizen.

Shipping the "CoviPen" along with the Covid-19 Home Vaccination Kits to all individuals in the USA, based on census data, will save citizens time, give them ease of application and offer another option to help them make the best decision for themselves and their families in this dire chapter in human history.

To add some control to this system, the kit will also arrive with a unique vaccination kit ID number that will be used by the VIN or individual applying the vaccine to log in to the covid19US.com database, using a login User Interface accessible on the Homepage. Once logged in, the user's internet browser will be directed to a citizen vaccination record page which will allow them to enter data into the database records. One line of input fields will be displayed for each member of the residence according to the most recent census. Additional lines may be added in case the number of individuals in the household has increased.

I will go into more detail about the database structure in the next chapter, but the record fields will include the individual's Social Security number (or an equivalent), the date they received the vaccine, which specific vaccine they received, and perhaps other useful fields like name, address, and contact information fields.

Not everyone will bother to fill out the database record form, and some might fill it out partially or incorrectly. However, as much as some of the people receiving the vaccine do give their information, it will help to document and track the results of this massive vaccination effort.

I am now going to address some obvious objections to this approach.

What about those who refuse to fill out the database record form on covid19US.gov? It doesn't matter. All that matters is that vaccine is getting into individuals.

Some people will receive a vaccine kit and refuse to take it. Sadly, this is their right, and I would not force it on people even if that were possible.

How about people who receive this free vaccine and then turn around and sell it for a profit? It doesn't matter. It is unlikely that many people would pay money for the vaccine only to throw it away. Most people who are motivated enough to pay money for it will do it because they want to be vaccinated or to have someone else vaccinated. If a vaccine is illicitly sold, a great majority of it will still be going into arms, and that is all that matters.

Another reasonable objection to this method is that it is probable that this approach will result in a maximum of waste of vaccine, and of monetary resources used to implement this system and get the vaccine to so many households.

Again, it doesn't matter. Does money or vaccine have a soul? No. Can they be reproduced or manufactured at will? Yes. As I stated earlier, this document is written from the point of view that the *only* consideration is saving human lives. To do this as quickly as possible, I am placing no importance on priorities associated with money or a possible waste of vaccine.

Despite the somewhat chaotic nature of this system, and the maximization of waste of money and vaccine, I strongly feel that this is the fastest way to get the entire population of any country vaccinated, and thereby contribute to saving the entire human species on this planet, before the COVID-19 and all of its variants get completely out of control.

Distribution System Two:
Also Fast, with More Control

The key to this system is the covid19US.gov (again, only a suggestion for the name) web site which will allow several key resources for the spreading of the cure to COVID-19.

As mentioned previously, a tutorial video educating those who elect to learn this technique will be available to anyone on the planet who desires to learn it.

In this system, instead of sending COVID-19 vaccine to every household in the USA, the virus will first be ordered by a certified VIN.

To become certified, an individual will be required to pass an exam that will demonstrate minimal competence in the procedure. Here is a suggested web site for the exam: http://www.covid19US.gov/VIN_final exam/

Once an individual has passed the exam, they are issued an VIN number, which is automatically entered into the Certified VIN field of their covid19us.gov database record. They will also be able to download a certificate of graduation that they will be able to print and frame. A copy of their certification can also be sent to their email address. They will now have the ability to log on using any computer on the planet and have vaccine vials or CoviPens shipped to them directly.

Upon successful passing of the final exam and the issuing of their VIN certificate, five to ten doses of vaccine will be automatically shipped to their address listed on their covid19us.gov database record. The certified VIN will be able to establish a login password which they will then use when they need to order more vaccine.

Example URL: http://www.covid19US.gov/cv19vaccine_orderform

Once a VIN has become certified, they should also be prepared and able to teach other people to give intramuscular inoculations, and to pass on the teaching.

One major advantage to both of these systems is that it allows families to get vaccinated without having to break quarantine. At least one member of every household should get certified as a VIN and will have the opportunity to order more vaccine for their community (or any place they are dispensing inoculations) as is needed.

Allowing families to get inoculated in their own homes allows them to avoid vaccine distribution sites which could have the tendency to become mass spreader events. Instead of lines of cars burning gas and polluting the environment, who sit in lines with their engines running and wasting money and gas simultaneously, they will be able to stay safely in their homes and get inoculated while saving time, money, fuel, and harm to the environment.

I believe that this approach will be less costly, far safer than making people leave their homes to get inoculated, and is environmentally friendly.

There will be some people who are not comfortable giving themselves an injection. Those people will have the vaccine in hand and can have it administered by another family member, a certified neighbor, go to any mass vaccination site, or even just go to a local medical center, pharmacy, or doctor's office.

The point here is to offer as many options as possible to those in need of vaccine inoculations. In an emergency, it is important to cultivate options. The more options offered to the public, the easier it will be to get everyone vaccinated as soon as possible.

New Inoculation Technologies

I am not going to spend much time on this section. It is more for future pandemics than it is for the current state of the world in relation to the COVID-19 virus and its variants.

DRONES: One transportation and delivery technology that might be of great help is flying drones. This technology is just starting to be implemented and used in actual industrial and business applications. UPS for one is now starting to actually use drones to enhance their ability to deliver parcels promptly. With little additional development, drone technology could help get vaccines directly delivered to American households.

Vaccination Booths: This concept will require some research and development to get a product that can be distributed on a large scale in our culture.

The design as I envision it would be a booth available in various venues such as grocery stores, pharmacies, and even town halls that offer vaccine injections for those who want them. An individual will be able to first enter the booth and sit down, and present their ID to be scanned for documentation purposes. They will also have the opportunity to present their insurance card, and a slot will be provided for reading a credit card for locations that might require a co-pay and/or application fees.

A computer screen will then allow the customer to select the injections they want to receive. The selection might include "Influenza", "Pneumonia", and "COVID-19", among others that can be added in the future as the need arises. When their selection is made, the customer will select the "Start" button, after which an automated robotic arm will extend from the background machinery and apply first a local antiseptic such as isopropyl alcohol and then the vaccine(s) chosen. Before the customer leaves, a dispenser will offer a bandage that they can apply manually.

Although I have not heard of such technology being in existence at this time, I believe that it could be developed and a final product released quite quickly if the motivation to do it is sufficient. One approach would be to assemble a group of American doctors and engineers working in cooperation with a Japanese robotics corporation. With a sufficient budget, such vaccination booths could be available in as little as three months. This timeline would deliver this technology in time to help inoculate the country from the current COVID-19 pandemic.

Vaccination Drones: This approach is farther from our immediate grasp, and would take longer to develop from concept to distributable final product.

The idea, in this case, would be that a flying drone could be developed that would contain a reservoir of vaccine that would be delivered door to door. When the drone rings the doorbell of a household, the residents would have the option of receiving an inoculation if they desire it, or pass and let the drone continue to the next address.

This idea would take the longest time and the most money to implement than the previous two. However, if the priority is escalated, a working version of such vaccination drones could be available to be rolled out in about one year. If such technology were to be developed now, it will be available to help defeat future pandemics quickly and efficiently.

One nation working together in a concerted massive effort: I was not going to address manufacturing at all, but I heard Mayor de Blasio of New York suggest that all potential chemical and bioengineering companies join in to develop versions of the vaccine that are already approved. I think that this is an outstanding idea. To apply it to my delivery system, the Johnson & Johnson vaccine would be ideal. If the President were to buy a couple hundred million doses or so and delegate them to be put into COVID-19 Home Vaccination Kits and shipped directly to American households in a desperate and sincere effort to inoculate our nation's population to the level of Herd Immunity and farther, this would probably be the fastest way to achieve this goal.

Phase Three: Database and Internet Web Site

Although to merely get as many people inoculated as possible, a database to track the distribution would not be required, it would be a great improvement and enhancement.

In fact, it is a reasonable concept that it would be unacceptable to perform a massive inoculation of a population without some attempt to track the process, specifically and most importantly where the vaccine is being successfully applied, and who has received their vaccination.

In this chapter, I am giving an example of what the database might look like, and how it will become important in expanding our inoculation process to become a world-wide effort, which is the subject of the following chapter, "Globalization and Worldwide Cooperation".

An exact model of the database along with a working prototype will be easy to develop in a day employing the effort of two highly competent database designers.

Here is a crude prototype I am providing as an example:

MAIN_TABLE

Social Security Number	Name	Inoculation 1 Date	Vaccine 1 Manfacturer
123-45-6789	John Doe	2/27/2021	J&J

AVIN Certification #	Household Vaccine Kit #	Mailing Address	Email Address	Mobile Phone
80176DD2021	HVK55556667789	100 Smith Ave	Jdoe@someurl.com	111-000-1234

The Social Security number would be declared as type integer in the data structure and serve as the main database record key for indexing needed for record search purposes. The secondary indexing key will be the VIN Certification Number, which will become important in Phase 4, as it will assist in contacting potential nurses via text message when a call goes out requesting that nurses go to a COVID-19 hotspot.

Databases frequently contain more than one table. Another helpful table would be a VACCINE_LOCATION table, which would contain records that have the location, quantity, and manufacturer of the vaccine stores and availability worldwide.

The web site will offer a page where graphs and maps of the distribution of both virus and vaccine will be made available.

A second page will offer video tutorials of how to administer the vaccination in various languages.

The site will include a final exam (mentioned previously) for those who desire to be certified as a VIN. Upon passing the exam, the newly certified nurse will be assigned a VIN Certification Number which will be automatically entered into the "VIN Certification #" field of their personal record, associated with their Social Security number, issued a graduation certificate, and allowed to enter a password that will identify their official nurse's account, from which vaccine can be ordered to be shipped to them directly.

To speed up the dissemination of the vaccine, several doses could be automatically shipped to the VIN's address on record. Additionally, the nurse will be able to modify this order and/or order more vaccine as needed.

All nurses giving a vaccine should use a computer interface to update the site's database for every individual they inoculate, and for each person whom they teach to administer vaccinations as well. Being able to track both of these data points will become essential to globalizing this system.

Implementing the system would also be inexpensive in terms of the Federal Governments' budget. To implement this system, one would need to hire a team that would require at least one software engineer, a database programmer, a web site designer, a systems administrator, and a healthy investment in web server hardware.

Defeating the virus in one country alone is nowhere near a solution to rid the world of COVID-19.

I suggest creating a template of the USA COVID-19 website and database which should be shared freely with every country capable of implementing this technology. They could use our template and customize it with their own language, traditional style, and flourishes. A simultaneous worldwide effort to defeat this pandemic will be required for the human species to be victorious.

Phase Four: Globalization
And Worldwide Cooperation

America First

Although there are reasonable philosophical reasons why it is morally better for us to service the countries in the greatest need first, I believe that getting the USA to Herd Immunity should be prioritized for the same reasons that we prioritize vaccinating doctors and other front-line medical personnel.

Once other countries start implementing our database and web site system, the sites' software will begin communicating with each other and keeping an up-to-date version of the world-wide database current on each server.

The COVID-19 servers in each country will poll all other worldwide servers every two minutes (a norm for email servers), download the current version of each country's database, and be capable of reflecting the exact state of the entire world in terms of vaccination progress, exactly which citizens have been vaccinated, and availability of vaccine and nurses.

As nurses vaccinate and train more nurses in performing vaccinations, they should update the database using an app or their cell phone or tablet, laptop, or another computer to access the country's COVID-19 web site from which additional vaccines will be requested and dispensed.

I envision a COVID-19 Analysis and Communications Utility Application that will take data from the CDC, Johns Hopkins, and WHO servers that track the infected and morbidity numbers related to the virus, to determine which nations are most in need of help. This COVID-19 Monitoring Utility program will also poll the COVID-19.gov server database to collate data on the availability of VIN nurses and vaccines.

The COVID-19 utility program will keep a watchful eye on the status of every country on Earth twenty-four hours a day, and automatically detect the locations most in need of the distribution of the vaccine. It will then be able to send out messages to VIN's, based on their location, using their contact information and VIN Certification number. Participation would be voluntary, and as more nations reach Herd Immunity more nurses will be available to help out regions in need. The program will also automatically send in an order to the location that has the largest excess of vaccine and have it express shipped to meet the nurses when they arrive on the scene of the outbreak.

The nurse's role will be powerful, but brief, according to my world-wide distribution plan. I want to describe an example to illustrate this point.

In this example, a nurse from North America flies to a small South American country in need, such as Bolivia.

When the nurse arrives and gets set up to inoculate members of the country's population, they will do more than just administer injections. They will be given the goal to give 100 individuals inoculations before heading home. They will also, and this is the key, teach every one of the people they vaccinate *how* to administer vaccinations, and update the person's COVID-19 database record to show that they are flagged as a nurse, and get an VIN Certification

Number assigned to them, and give each of them enough vaccine to then train and vaccinate ten more individuals. Each of those people will then be instructed to order vaccine through the COVID-19.gov web site for their country, and be tasked to vaccinate and train ten more individuals. They will have the option of choosing the format of the vaccine, whether in standard vials or the new CoviPen syringe, depending on which is the most effective and helpful given the context and environment in which the vaccines will be administered.

This approach totally changes the math in terms of how fast the country might be able to reach Herd Immunity. If the nurse had stayed long enough, they might have inoculated another 100 people. However, using the system of distributing vaccines and the technical training needed to perform vaccinations, the cure for the COVID-19 virus can spread exponentially, just as the virus does when it reproduces.

So, to follow the example through, and to calculate the effect over several days, a nurse not training people to pass on the technique, who inoculates a hundred people a day, will have done 300 individuals in 3 days.

However, following my guidelines the nurse using my system will have been responsible for the vaccination of 1100 individuals after the second day has ended. That includes the 100 people they vaccinated and trained personally, added to the 1000 vaccinations which they sent out with the newly trained nurses (ten per new nurse, 100 vaccinated nurses). Once each one of the people in the next level down from the nurse have trained and vaccinated 10 additional people (which can easily be done in a day), then there will be 11,000 more people capable of administering the vaccine just two days after the VIN who came from North America had administered the first set of vaccines and trained the first 100 new VIN's.

Note that on the fourth day after the VIN's arrival there will be 110,000 new VIN's in the country, all certified to order vaccine from the COVID-19 website so that the process can continue. As long as each new VIN can train ten more in a day, by the fifth day there will be 1,100,000 new vaccine nurses due to the efforts of just one volunteer from another country. Two days after that, the country could have up to One Hundred and Ten Million newly trained vaccination nurses, and the vaccine is being airlifted in daily at this point.

Now, what if ten or even a hundred nurses hail the call when the request comes in via message? Nurses could also potentially come in from the U.K., France, Italy, Canada, and who knows how many other countries? When the nations of the world work together, using this system, countries in need will be able to get vaccinated at truly astonishing speeds!

Final Note:
Speed and Urgency

The fact that the COVID-19 virus has already been able to mutate and evolve into at least three other strains that are more suitably adapted to survive and more capable of harming humans is a bad sign. The reason this pandemic needs to be eradicated quickly is that we can't afford for the new variants to continue to evolve to be heartier, more resistant, more contagious, and more virulent.

This is why cost and waste of reproducible resources is irrelevant.

Only the number of human lives saved matters at all at this point. I beg all world leaders who read my plan to implement it, or a variation of it, and get their country's population to Herd Immunity as soon as possible, regardless of cost and waste.

Appendix A: Remuneration to Schpleee Technologies, Inc.

For payment to Schpleee Technologies, Inc. for this Expedited Inoculation System, I ask for $0.00.

What I do want is for you, the reader, to do everything you can to save as many lives as you can. If you are not a politician, please send this to your local and federal representatives and senators. I ask that all people of all nations bring this plan to your leaders. I ask that all human inhabitants of the Earth work together in unity to defeat this common enemy, which is a threat to all of us, and everyone we love.

www.ingramcontent.com/pod-product-compliance
Lightning Source LLC
Chambersburg PA
CBHW070511220526
45467CB00002B/615